Rockjam 88 Keys Keyboard

User Manual

The all-inclusive user manual to guide you through.

Contents

DESCRIPTION

The RockJam 88 digital piano keyboard comes with an 88 full-sized, semi-weighted, velocity-sensitive keys that directly replicate the feel of a real piano.

This weighted piano keyboard is filled with current features which includes ten unique voices such as: Hammond organ electric keyboard piano, strings, synth, upright piano, grand piano, bass, guitar, percussion, and church organ.

It has inbuilt stereo speakers on its electric keyboard which produces a powerful 24 Watts of sound.

This full-scale keyboard piano has a headphone input so you can have your private practice without disturbing others.

This piano keyboard comes with weighted a key which includes a USB input which enables you to play along with your favorite songs.

this piano keyboard has impute which include; ¼ Inch (6.35 mm) sustained pedal input, ¼ Inch (6.35 mm) soft-pedal input (pedal not included), 7 pin sostenuto pedal input and ¼ Inch (6.35 mm) microphone connector. Note: the pedals and microphone are not included.

The Piano keyboard outputs; ¼ Inch (6.35 mm) stereo headphone output for private practice, and stereo Aux output to connect to an external recording desk, mixing device, amplifier, or sound system.

You will have exclusive access to content inside the Simply piano application for iOS and Android devices to help in learning piano for beginners.

the keyboard piano package includes Keynote stickers that are very easy to fit, it gives you a visual signal of the right keys to play.

So, go ahead and click on the 'buy now' Button to get started. The details below must be followed to the letter:

Adapter for power:

- Please make use of only the AC adapter that came with the kit.
- The electronic keyboard may be affected by an incorrect or defective adapter.
- Avoid positioning the power cord or AC adapter near any heat source, like heaters or radiators.
- Please make sure that no large items are put on the power cord and that it is not subjected to stress or over bending to avoid damaging it.
- Make sure the power plug is clean and clear of surface dirt regularly. Wet hands should not be used to plug in or unplug the power cable.

Use of the electronic keyboard:

Don't open the electronic keyboard's body:

- Do not attempt to open or dismantle the electronic keyboard. Please stop using the device and send it to a trained service agent for repair if it is not working properly. The use of an electronic keyboard is as follows:

- Please do not position the electronic keyboard in a dusty setting, direct sunlight, or in areas where the temperature is extremely high or extremely low to avoid damaging the appearance or internal parts.

- Avoid using the electronic keyboard on a slanted surface. Avoid placing any liquid-filled vessel on the electronic keyboard to avoid spillage thereby damaging the internal components.

Maintenance:

Wipe the electronic keyboard's body with a dry, soft cloth to clean it.

Connection:

Please set the volume of any peripheral device to the lowest setting to avoid damage to the electronic keyboard's speaker. Then slowly raise the volume to an acceptable level as the music begins to play.

While operating it:

- Do not make use of the keyboard at its full volume for an extended period.
- Do not put heavy items on the keyboard or use unnecessary force when pressing the keys.

- Only a responsible adult should open the package, and any plastic packaging should be properly preserved or disposed of.

Specification

The specification can be changed without any notice.

Indicators, Controls, and External Connections

Front Panel

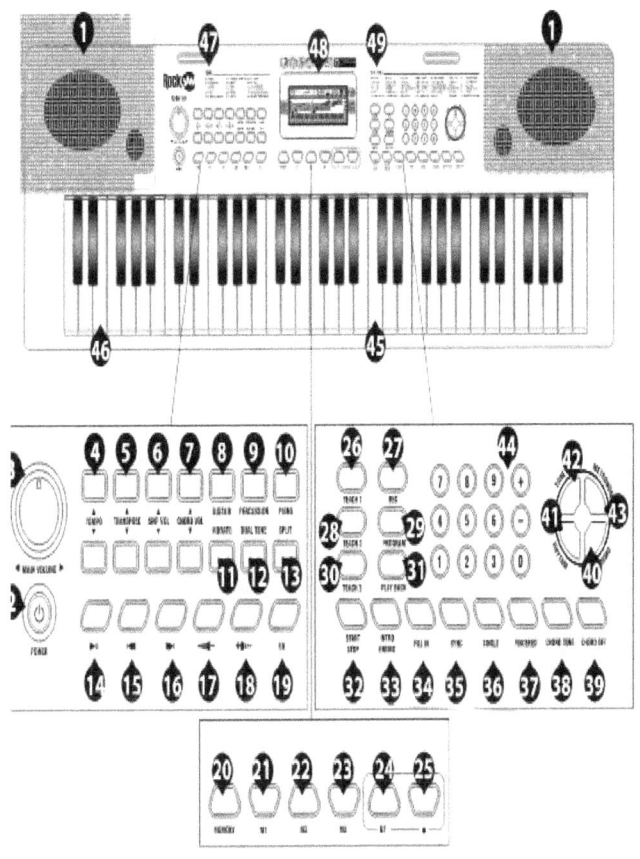

1. Stereo Speakers
2. Power Switch
3. Sync

21. Memory Storage 1
22. Memory Storage 2
23. Percussion

4. Single Finger Chords
5. Fingered Chords
6. Fill In
7. Metronome
8. Split Keyboard
9. Vibrato
10. Start / Stop
11. Intro / Ending
12. Main Volume +/-
13. Tempo [Fast/Slow]
14. Accompaniment Volume +/-
15. Transpose
16. Sustain
17. Record
18. Rhythm Program
19. Playback
20.Memory Function

24. Play/ Pause
25. Previous Track
26. Next Track
27. Music Volume -
28. Music Volume +
29. Number Pad
30. Tone
31. Rhythm
32. Demo
33. Teach 1 and 2
34. Rhythms List
35. LED Display
36. Tones List
37. Chord Keyboard Area
38. Keyboard Playing Area

External Connections

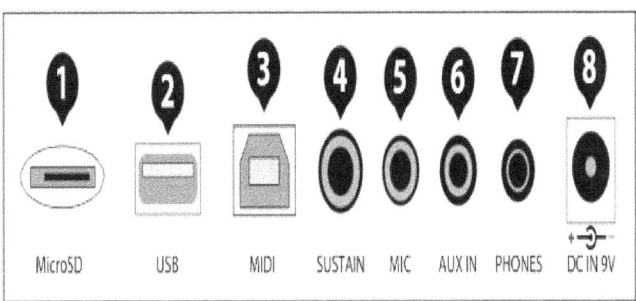

1. TF Card Slot

2. USBMP3 Input

3. MIDI Input/output

4. Sustain Pedal Input

5. MIC Input

6. AUX-IN

7. Headphone Output

8. DC IN

LCD Display

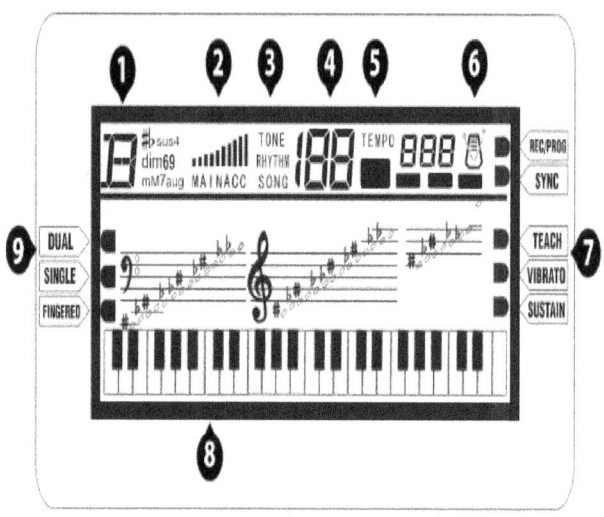

1. Playing chord indication

2. Volume level display

3. TONE, RHYTHM, or DEMO song

4. Digital display for song, rhythm, and tone number

5. Tempo value of rhythm and song (bpm)

6. Metronome for rhythm and song

7. Mode Indicators 1

8. Key Indicator

9. Mode Indicators

Preparation before the First Usage

Power

Use of the AC/DC power adapter:

Please make use of the AC/DC power adapter coming alongside the electronic keyboard. You can also make use of a power adapter having a center-positive plug with a 1,000mA output current and a DC 12V output voltage. Link the power adapter's DC plug to the DC12V power socket on the back of the keyboard, and then connect the power cord and switch it on.

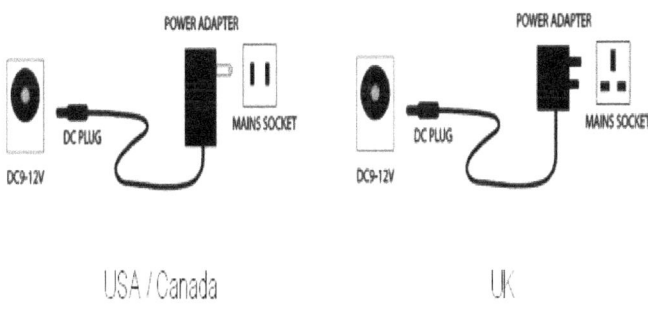

USA / Canada   UK

USA / Canada UK Caution:

Using the battery:

Place 6x 1.5V Size D alkaline batteries into the battery lid on the bottom of the electronic keyboard. Replace the battery lid after ensuring that the batteries are inserted with the right polarity.

WARNING:

It is not a good idea to combine old and new batteries. If the keyboard is not in use, do not leave batteries in it. This would prevent any potential damage arising from leaking batteries. When the keyboard is not in use, disconnect the power adapter from the socket.

Accessories and Jacks

Using headphones:

Connect the 3.5mm headphone jack on the back of the keyboard to the phone jack. When headphones are attached, the internal speaker will automatically turn off.

Also included are the headphones with volume control on their cables.

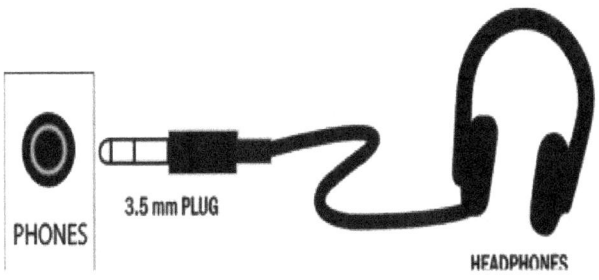

PHONES 3.5 mm PLUG HEADPHONES

Connecting a Hi-Fi Device or an Amplifier:

The built-in speaker system on this electronic keyboard can be connected to external hi-fi equipment or amplifier or other hi-fi equipment. To connect, switch off the power to the keyboard as well as any external equipment you choose to attach. Next, link one end of a stereo audio cable (not included) to the AUX IN or LINE IN socket on the external device and the other end to the phone jack located at the electronic keyboard's back.

Sustain Pedal Connection

A 6.35mm port on this electronic keyboard can be used to attach a sustain pedal (not included). Attach the 6.35mm socket to the sustain pedal.

Connecting an Android device or iPad:

Through the MIDI output, the Keyboard can be attached to an Android or Apple device.

This lets you listen to music via an app. Connect the USB B type plug to the keyboard's MIDI output at the back. The USB cables for Apple and Android are not included but can be purchased on eBay or Amazon.

Note thatthe MIDI feature requires Android 6.0 or higher.

The Recommended app: Simply Piano by JoyTunes, available on the Apple App Store and Google Play Store.

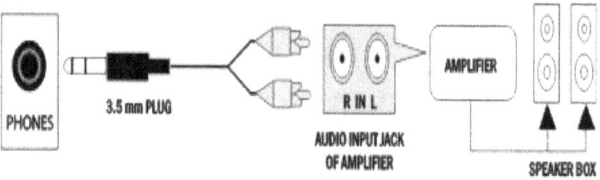

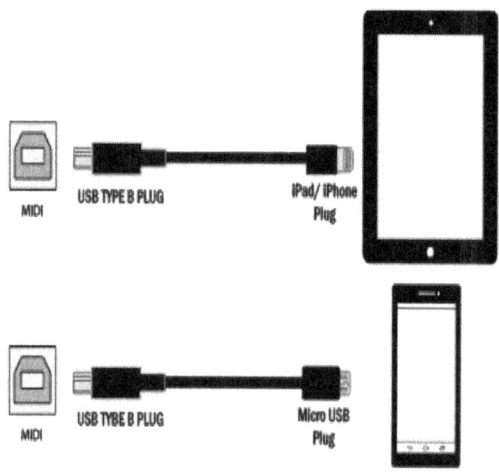

To play music via the keyboard, connect a phone or audio device to the AUX Input:

This keyboard has a built-in speaker system that can play music from your phone or other mobile devices. Connect the other end of a stereo audio cable to your phone or audio device and plug it into the AUX IN socket on the back of the keyboard.

Check to see if the keyboard is turned on.

To adjust the volume of the song, use the phone's volume control.

The AUX IN cable is not included.

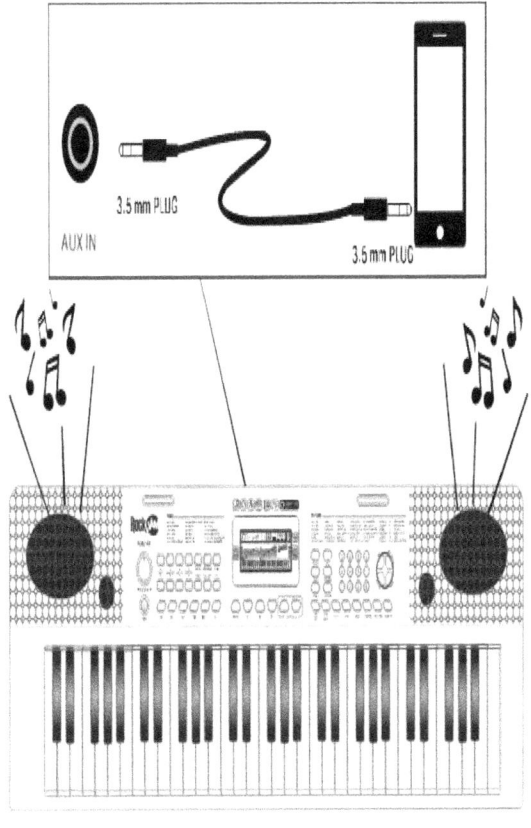

PC or Laptop Connection:

You can use third-party MIDIsoftware to record and playback songs via the midi files by connecting the keyboard to a PC or laptop. Attach a USB B type cable to the MIDI output on the keyboard's back, then connect the USB type A cable to the laptop or PC.
The USB cable for connecting to a computer or laptop is not included.

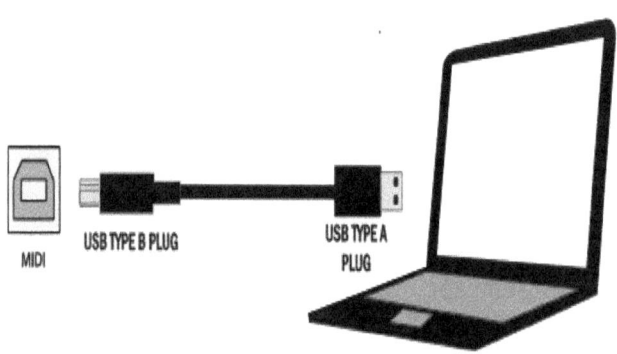

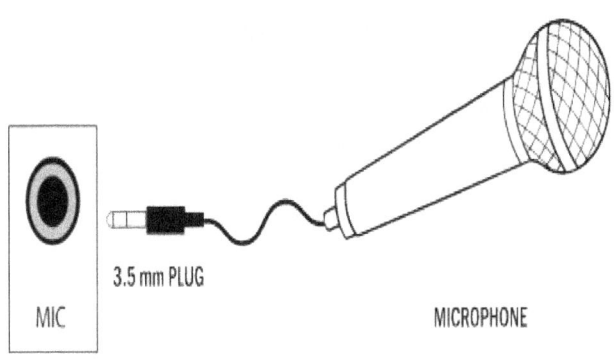

3.5 mm PLUG

MIC

MICROPHONE

Microphone Connection:

Connect the 3.5mm microphone jack to the MIC jack that is located at the keyboard's rear.

Connecting a TF card or USB flash drive:

Place files on a TF card or USB flash drive to play MP3s via the Keyboard Speakers.

The first song will play after you insert the card into the input slot on the back of the keyboard.

• To play or pause playback, click the button.

• To go to the previous or next song, press the or keys.

• To raise or decrease the volume of the playback, click the or keys.

• Press the button to change the playback tone. There are six different tone settings, and each press will change one of them.

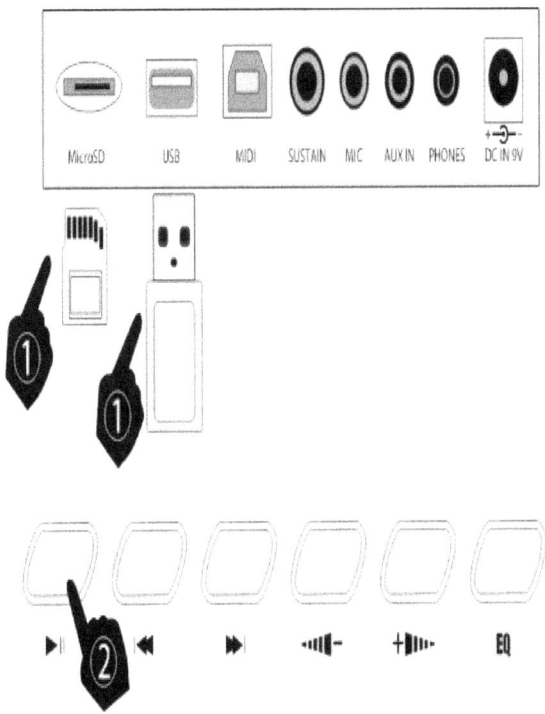

MicroSD USB MIDI SUSTAIN MIC AUX IN PHONES DC IN 9V

Amazon Alexa Operation Setup

You'll need the following to set up the keyboard for use with Amazon Alexa:

•An Android or Apple iOS smartphone or tablet (iPhone, iPad, etc.)
•An internet connection
•An Amazon account that is active

The Big Picture

Three steps must be completed to use Amazon Alexa:
•On your smartphone or tablet, download and install the Alexa companion APP "RockJam Keyboard."
•Connect the keyboard to your WiFi network with the aid of the companion app.
•Log into your Amazon account and register the keyboard. Additionally, you should install the following Alexa skill to assist you in learning to play:

•Download and install the Amazon Alexa app on your smartphone.

Look for the "Piano Teacher skill" and activate it.

The detailed essential steps for setup (WiFi Setup and RockJam APP)

On your smartphone or tablet, download and install the Alexa companion app "RockJam Keyboard" from eitherthe Apple App Store (iOS) (iPhone, iPad) or the Google Play Store (Android).

When the keyboard is turned on for the first time, it should enter the network communication mode. If it does not happen, or if you want to use the keyboard with a different WiFi connection, follow the steps below to set up a network connection.

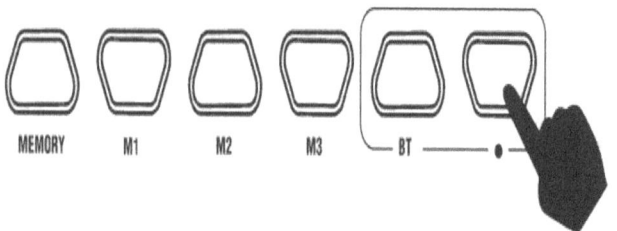

To enter the network connection mode, press down the blue
Alexa button for 5 seconds as displayed.

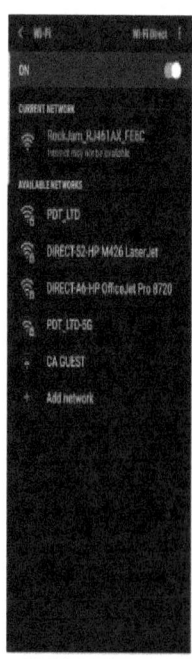

A notification will be displayed thus, "Entering Network Connection Mode"

When you open the "RockJam Keyboard" app on your smartphone, you will be prompted to connect to the keyboard's WiFi network.

Link to the keyboard WiFi network, RockJam RJ461AX XXXX, using your smart device's WiFi settings.

If the WiFi doesn't work, go back to the "RockJam Keyboard" app and select "Could not locate RockJam RJ461AX XXXX?"Then try again, following the directions on the screen.

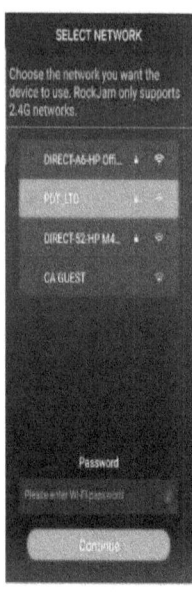  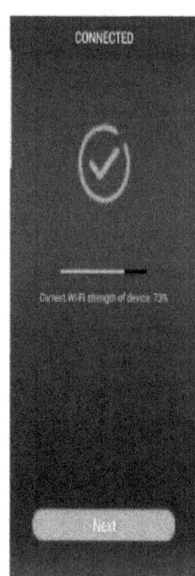

After that, you'll be prompted to pick your home WiFi network and input your WiFi passcode.It's worth noting that you can only use a 2.4 GHz WiFiconnection.

The keyboard will connect to your home WiFi after you press Next. After that, you'll be prompted to connect your keyboard to an Amazon account.

Note that if you miss this stage, you will not be able to use Alexa with the keyboard.

After that, your smart device will reconnect to your home's WiFi network. Both your keyboard and the smart device will be connected to your home WiFi at this stage.

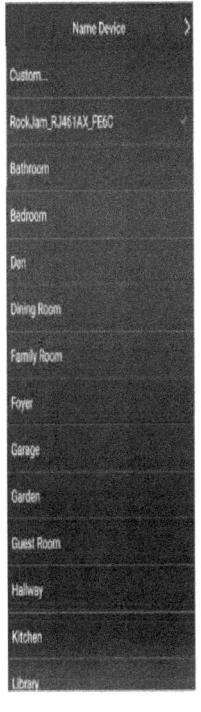
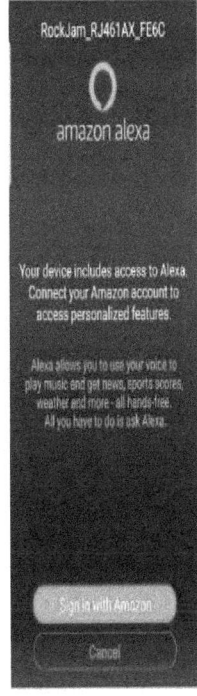

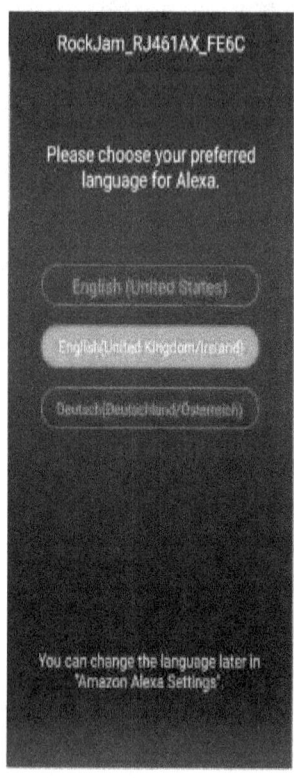

By asking Alexa a simple question, you will see if the keyboard has been successfully registered. Say something like, "What is the time?" after pressing the blue Alexa button on the keyboard. Alexa's answer should be the current time.

Appendix VII contains a complete list of the additional features provided by the "RockJam Keyboard" companion app.

Operations of the Keyboard Volume and Power

Controlling power:

To turn the switch on, press the [POWER] button; to turn it off, press the [POWER] button once more.

When the power is switched on, the LED screen will light up (Wi-Fi is switched on).

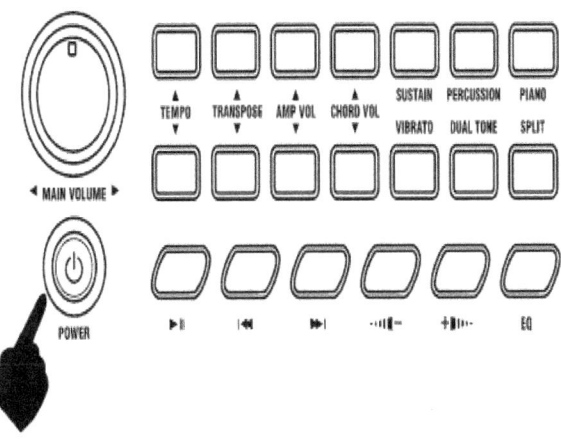

Adjustment of the Master Volume:

Whirl the MAIN VOLUME dial to change the volume.

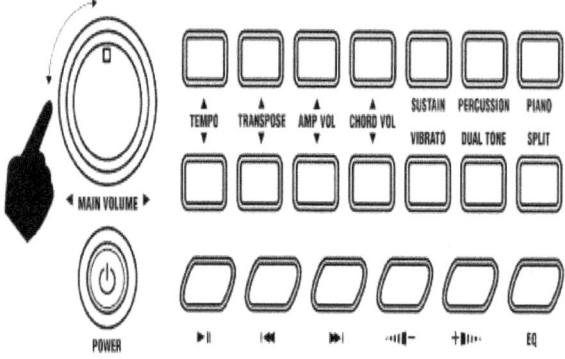

Tone

Tone Selection:

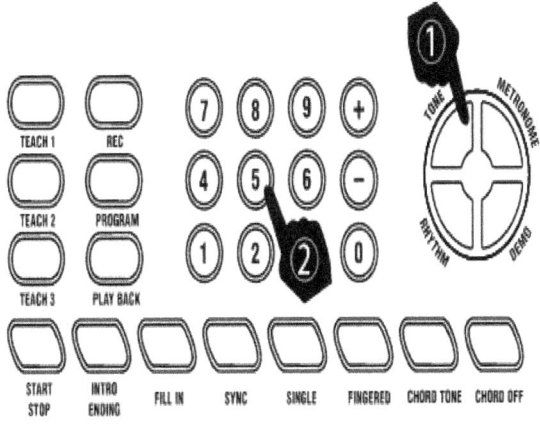

The default tone upon switching on the keyboard is "00 Grand Piano." To adjust the tone, click the TONE button first, then press the corresponding digits 0-9. The [+ / -] buttons on the keypad can also be used to change tones. For a complete list of available tones and codes, see Appendix III.

Effect &Control

Dual Tone Keyboard:

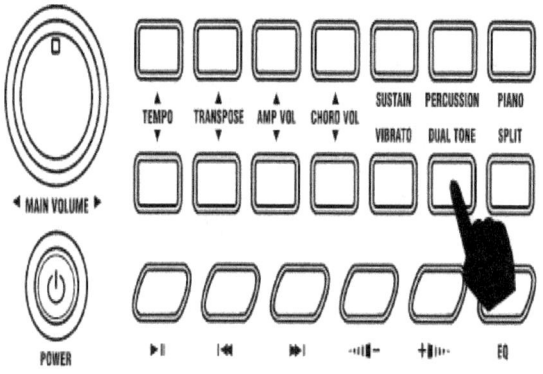

The DualTone mode lets the keyboard mix two tones and plays them together. To activate the Dual Tone Keyboard mode, press the [DUAL ONE] button. A flag indicator will appear on the LCD to indicate that the Dual Tone mode is active.

The first tone will remain the same as the tone selected before pressing the [DUALTONE] button. The second tone can then be freely chosen by pressing the related digits on the keypad. To exit DualTone mode, press the [DUALTONE] button once more.

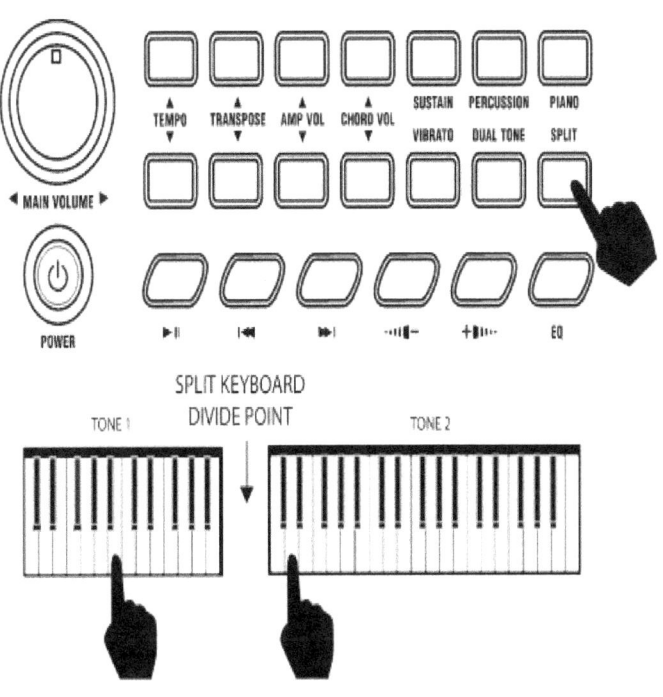

Split Keyboard:

Press the [SPLIT] button to activate Split Keyboard mode. At the 24th key from the left, the keyboard splits into two keyboards. In the Split Keyboard mode, the left-hand keys' pitch will be raised by an octave. To exit Split Keyboard mode, press the [SPLIT] button once more.

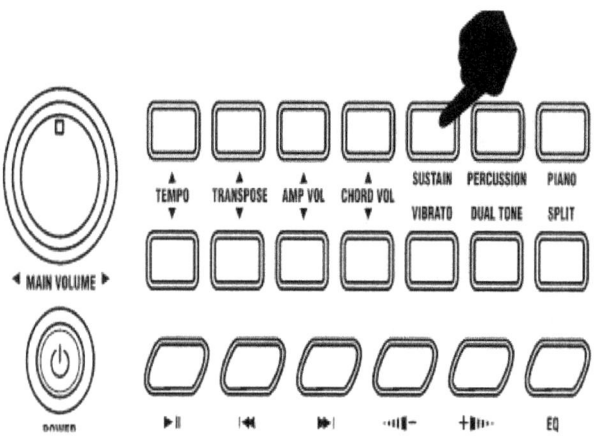

Sustain:

To enter Sustain mode, press the [SUSTAIN] button.

[SUSTAIN] will be shown on the LCD. When you choose this mode, the sound of each note you play is lengthened. The [SUSTAIN]button must be pressed again to turn off the sustain function and exit this mode.

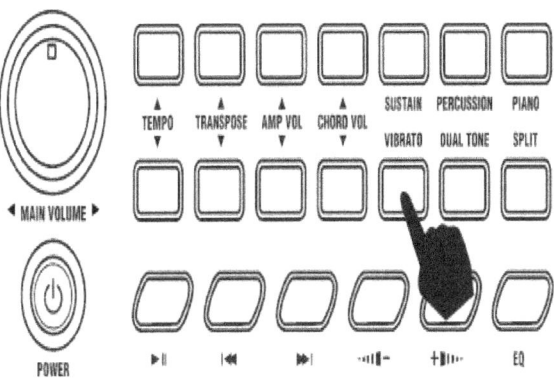

Vibrato (Vibrato):

To enter Vibrato mode, press the [VIBRATO] button. The [VIBRATO] will be displayed on the LCD.

When this mode is triggered, a trembling effect is applied to the end of each note played.

The Vibrato function will be turned off and this mode will be exited by pressing the [VIBRATO] button again.

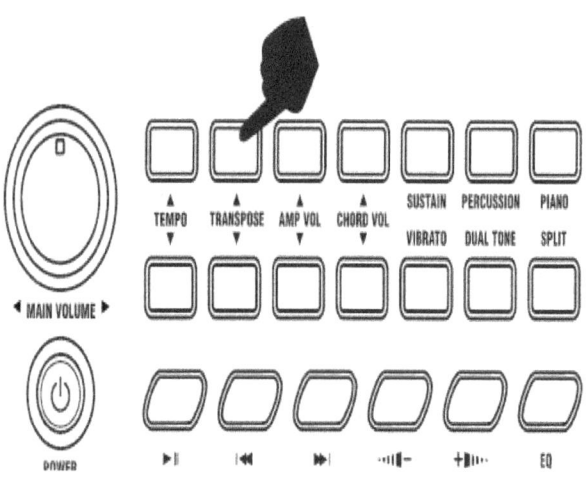

Transpose:

To adjust the musical scale of the note being played, click the [TRANSPOSE] keys. The scale can be adjusted by 6 steps upwards or downwards. The musical scale will be reset to oo by pressing both [TRANSPOSE] buttons at the same time. After powering on and off, the transpose level is reset to oo.

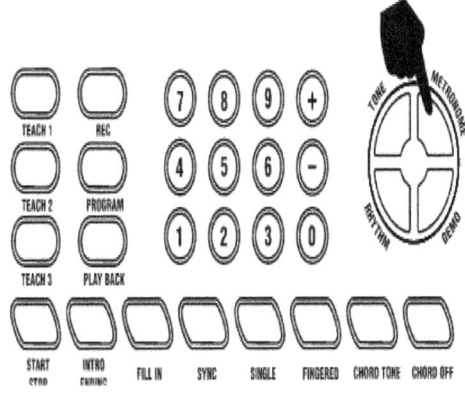

Metronome:

To begin the tick-tock beat, press the [METRONOME] button. You can choose from four different beats. Tap the [TEMPO] buttons to speed up or slow down the rhythm, depending on what is needed.

To peruse through the available beat patterns, repeatedly press the [METRONOME] button.The beat that you've selected will be shown on the LCD monitor.

When you begin playing, the metronome effect is applied to the song.
Touch the [METRONOME] button again or the [START/STOP] button to exit this mode.

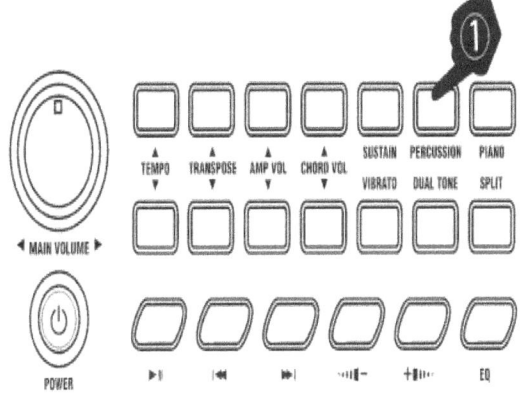

PERCUSSION KEYBOARD AREA

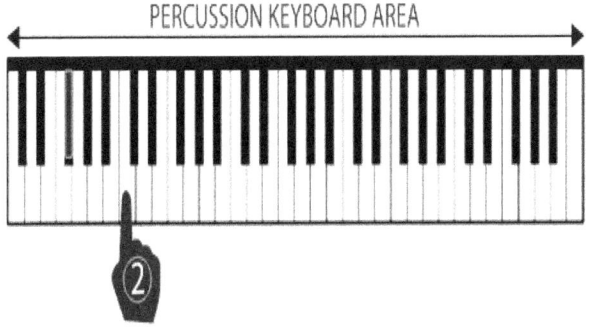

Panel Percussion Instruments:

Any of the keyboard's keys can be used to play percussion sounds when the [PERCUSSION] button is pressed.

To exit Percussion mode, press the [PERCUSSION] button once more. Appendix I is a good place to start for a list of the 61 available percussion sounds.

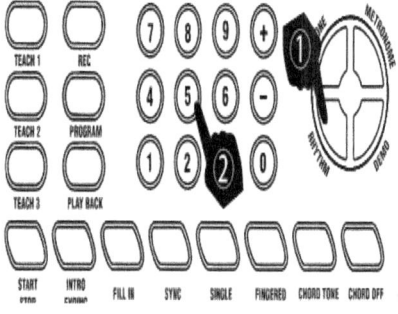

Rhythm:

Choosing the rhythm:

Any of the 200 built-in rhythms can be selected.

Please see Appendix II for more information on the rhythm table

To access the rhythm selection feature, press the [RHYTHM] button. The present rhythm number will be shown on the LCD monitor.

You can choose the rhythm you like by pressing the +/- buttons or the corresponding digits on the numerical keypad.

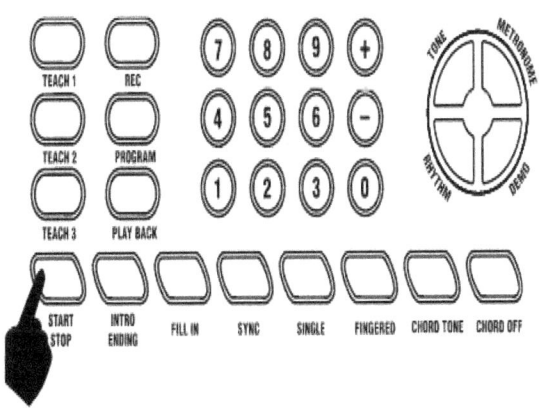

Start / Stop:

After choosing a rhythm, you can press the [Start/Stop] button to either start or stop playing the rhythm that you have chosen.

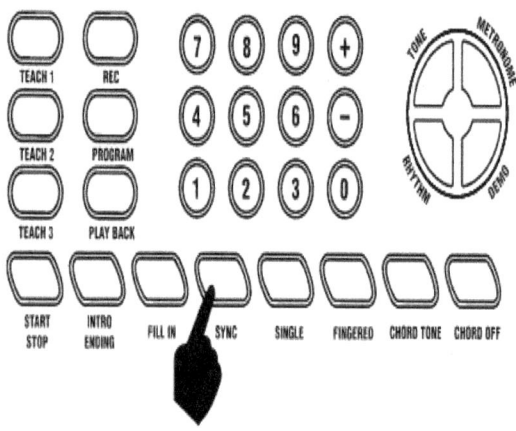

Sync:

Press the [SYNC] button to activate the related mode. Then, on the left-hand side of the keyboard, press any of the first 19 keys to begin playing the selected rhythm.

To stop the rhythm and exit the sync function, press the [START/STOP] button.

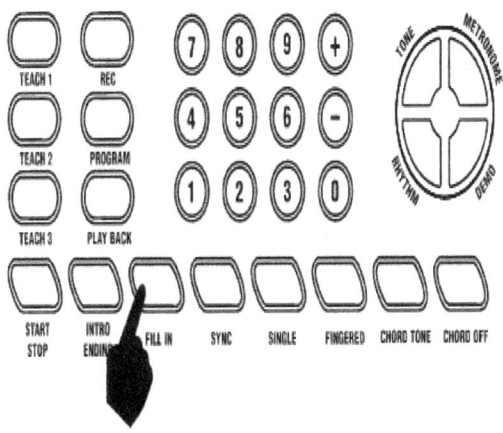

Fill in:

You can include a rhythmic fill as you play a rhythm by pressing the [FILLIN] button during playback. The rhythm will carry in playing as normal after the fill-in.

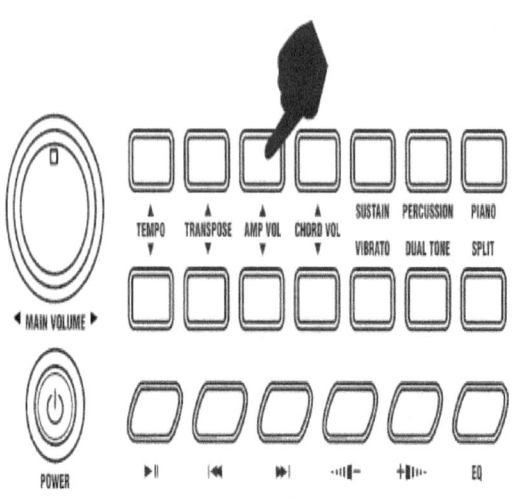

Accompaniment Volume Adjustment:

Press the [ACCOMPVOLUME] buttons to adjust the
Accompaniment volume. The volume will be displayed on the
LCD as the volume is adjusted. There are 10 levels of the
adjustment range, which are displayed on the LCD as a bar
graph.

Pressing both [ACCOMPVOLUME] buttons at the same time
will reset the Accompaniment Volume to its default setting

(level 006). Notice that the Main Volume control affects the accompaniment's output level as well.

The accompaniment volume will be reset to the default level when the power is switched on.

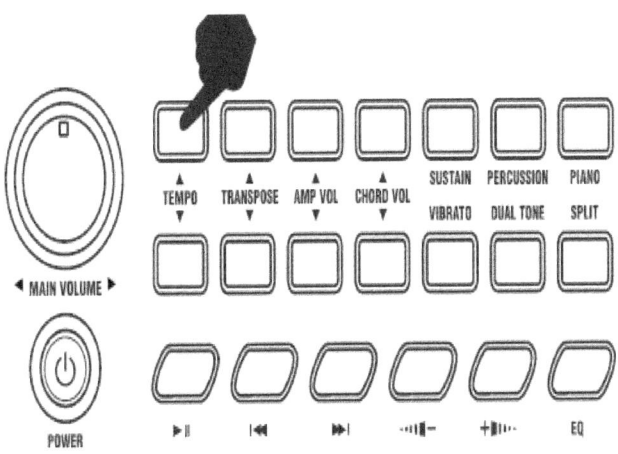

Changing the tempo:

To adjust the metronome, demo songs, or tempo of the beat, press the [TEMPO] keys.

The range of change is 30-240 bpm. If you click both [TEMPO] buttons at the same time, the tempo will be reset to its default. The tempo will be 120 bpm when the power is turned on.

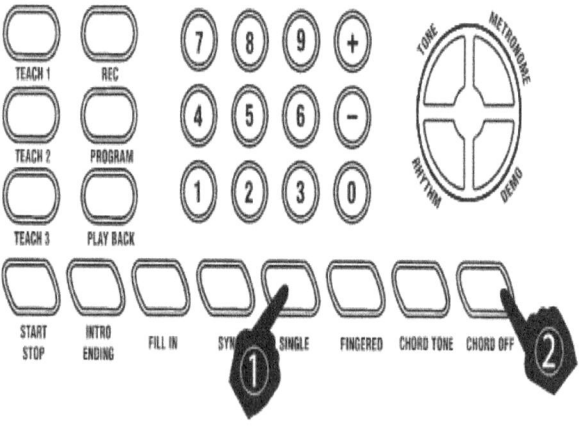

Chords Accompaniment

Chords with just one finger:

To use the single finger chord function, press the [SINGLE] button. The LCD screen will show which chords are activated. While a rhythm is playing, chords are played by pressing a key in the chord area on the keyboard's left-hand side (keys 1-19). You can find the required finger patterns in Appendix VI. The chord that is being played is shown in the LCD's top left corner.

To stop the chord accompaniment, press the [START/STOP] button.

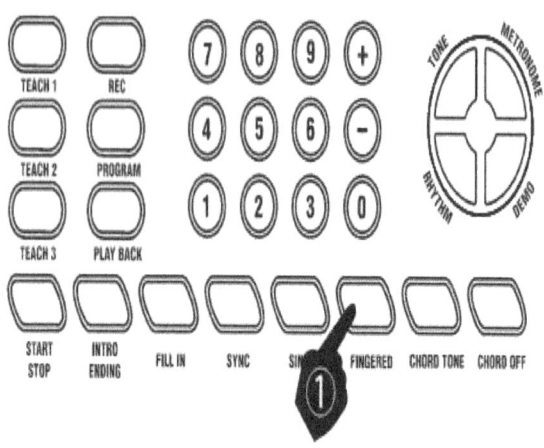

Multi-Finger Chords:

To use the multi-finger chord feature, press the [FINGERED] button. Which feature is active will be shown on the LCD screen. While a rhythm is playing, chords are played by pressing keys in the chord area on the keyboard's left side (keys 1-19). Appendix VI contains the necessary finger patterns. The play chord is shown in the LCD's top left corner.

To start or stop the chord accompaniment, press the [START/ STOP]button.

Note that it is vital to make use of the finger patterns shown for this purpose in Appendix VI.

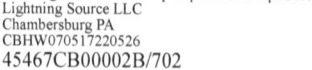